This planner belongs to :

_____

Twenty-four

January

S	M	T	W	T	F	S
	1	2	3	4	5	6
7	8	9	10	11	12	13
14	15	16	17	18	19	20
21	22	23	24	25	26	27
28	29	30	31			

February

S	M	T	W	T	F	S
				1	2	3
4	5	6	7	8	9	10
11	12	13	14	15	16	17
18	19	20	21	22	23	24
25	26	27	28	29		

March

S	M	T	W	T	F	S
					1	2
3	4	5	6	7	8	9
10	11	12	13	14	15	16
17	18	19	20	21	22	23
24	25	26	27	28	29	30
31						

April

S	M	T	W	T	F	S
	1	2	3	4	5	6
7	8	9	10	11	12	13
14	15	16	17	18	19	20
21	22	23	24	25	26	27
28	29	30				

May

S	M	T	W	T	F	S
			1	2	3	4
5	6	7	8	9	10	11
12	13	14	15	16	17	18
19	20	21	22	23	24	25
26	27	28	29	30	31	

June

S	M	T	W	T	F	S
						1
2	3	4	5	6	7	8
9	10	11	12	13	14	15
16	17	18	19	20	21	22
23	24	25	26	27	28	29
30						

July

S	M	T	W	T	F	S
	1	2	3	4	5	6
7	8	9	10	11	12	13
14	15	16	17	18	19	20
21	22	23	24	25	26	27
28	29	30	31			

August

S	M	T	W	T	F	S
				1	2	3
4	5	6	7	8	9	10
11	12	13	14	15	16	17
18	19	20	21	22	23	24
25	26	27	28	29	30	31

September

S	M	T	W	T	F	S
1	2	3	4	5	6	7
8	9	10	11	12	13	14
15	16	17	18	19	20	21
22	23	24	25	26	27	28
29	30					

October

S	M	T	W	T	F	S
		1	2	3	4	5
6	7	8	9	10	11	12
13	14	15	16	17	18	19
20	21	22	23	24	25	26
27	28	29	30	31		

November

S	M	T	W	T	F	S
					1	2
3	4	5	6	7	8	9
10	11	12	13	14	15	16
17	18	19	20	21	22	23
24	25	26	27	28	29	30

December

S	M	T	W	T	F	S
1	2	3	4	5	6	7
8	9	10	11	12	13	14
15	16	17	18	19	20	21
22	23	24	25	26	27	28
29	30	31				

Year in Pixels

	J	F	M	A	M	J	J	A	S	O	N	D
1.												
2.												
3.												
4.												
5.												
6.												
7.												
8.												
9.												
10.												
11.												
12.												
13.												
14.												
15.												
16.												
17.												
18.												
19.												
20.												
21.												
22.												
23.												
24.												
25.												
26.												
27.												
28.												
29.												
30.												
31.												

Color Codes

Notes

January 2024

MONDAY	TUESDAY	WEDNESDAY	THURSDAY
1	2	3	4
8	9	10	11
15	16	17	18
22	23	24	25
29	30	31	

January

2024

FRIDAY	SATURDAY	SUNDAY	NOTES
5	6	7	○
			○
			○
			○
12	13	14	○
			○
			○
			○
			○
19	20	21	○
			○
			○
			○
			○
26	27	28	○
			○
			○
			○
			○
			NOTES

February 2024

MONDAY	TUESDAY	WEDNESDAY	THURSDAY
			1
5	6	7	8
12	13	14	15
19	20	21	22
26	27	28	29

February

2024

FRIDAY	SATURDAY	SUNDAY	NOTES
2	3	4	○
			○
			○
			○
9	10	11	○
			○
			○
			○
			○
16	17	18	○
			○
			○
			○
			○
23	24	25	○
			○
			○
			○
			○
			NOTES

March

MONDAY	TUESDAY	WEDNESDAY	THURSDAY
4	5	6	7
11	12	13	14
18	19	20	21
25	26	27	28

March 2024

FRIDAY	SATURDAY	SUNDAY	NOTES
1	2	3	○
			○
			○
			○
8	9	10	○
			○
			○
			○
15	16	17	○
			○
			○
			○
			○
22	23	24	○
			○
			○
			○
			○
29	30	31	NOTES

April 2024

MONDAY	TUESDAY	WEDNESDAY	THURSDAY
1	2	3	4
8	9	10	11
15	16	17	18
22	23	24	25
29	30		

April

2024

FRIDAY	SATURDAY	SUNDAY	NOTES
5	6	7	○
			○
			○
			○
			○
12	13	14	○
			○
			○
			○
19	20	21	○
			○
			○
			○
			○
26	27	28	○
			○
			○
			○
			○
			NOTES

May

2024

MONDAY	TUESDAY	WEDNESDAY	THURSDAY
		1	2
6	7	8	9
13	14	15	16
20	21	22	23
27	28	29	30

May

FRIDAY	SATURDAY	SUNDAY	NOTES
3	4	5	○
			○
			○
			○
			○
10	11	12	○
			○
			○
			○
			○
17	18	19	○
			○
			○
			○
			○
24	25	26	○
			○
			○
			○
			○
31			NOTES

June
2024

MONDAY	TUESDAY	WEDNESDAY	THURSDAY
3	4	5	6
10	11	12	13
17	18	19	20
24	25	26	27

June

FRIDAY	SATURDAY	SUNDAY	NOTES
	1	2	○
			○
			○
			○
			○
7	8	9	○
			○
			○
			○
			○
14	15	16	○
			○
			○
			○
			○
21	22	23	○
			○
			○
			○
			○
28	29	30	NOTES

July

MONDAY	TUESDAY	WEDNESDAY	THURSDAY
1	2	3	4
8	9	10	11
15	16	17	18
22	23	24	25
29	30	31	

July

FRIDAY	SATURDAY	SUNDAY	NOTES
5	6	7	○
			○
			○
			○
			○
12	13	14	○
			○
			○
			○
			○
19	20	21	○
			○
			○
			○
			○
26	27	28	○
			○
			○
			○
			○
			NOTES

August

MONDAY	TUESDAY	WEDNESDAY	THURSDAY
			1
5	6	7	8
12	13	14	15
19	20	21	22
26	27	28	29

August

2024

FRIDAY	SATURDAY	SUNDAY	NOTES
2	3	4	○
			○
			○
			○
			○
9	10	11	○
			○
			○
			○
			○
16	17	18	○
			○
			○
			○
			○
23	24	25	○
			○
			○
			○
			○
30	31		NOTES

September

MONDAY	TUESDAY	WEDNESDAY	THURSDAY
2	3	4	5
9	10	11	12
16	17	18	19
23	24	25	26

September

FRIDAY	SATURDAY	SUNDAY	NOTES
		1	○
			○
			○
			○
			○
6	7	8	○
			○
			○
			○
13	14	15	○
			○
			○
			○
			○
20	21	22	○
			○
			○
			○
			○
27	28	29	30

October 2024

MONDAY	TUESDAY	WEDNESDAY	THURSDAY
	1	2	3
7	8	9	10
14	15	16	17
21	22	23	24
28	29	30	31

October 2024

FRIDAY	SATURDAY	SUNDAY	NOTES
4	5	6	○
			○
			○
			○
			○
11	12	13	○
			○
			○
			○
18	19	20	○
			○
			○
			○
			○
25	26	27	○
			○
			○
			○
			○
			NOTES

November

MONDAY	TUESDAY	WEDNESDAY	THURSDAY
4	5	6	7
11	12	13	14
18	19	20	21
25	26	27	28

November 2024

FRIDAY	SATURDAY	SUNDAY	NOTES
1	2	3	○
			○
			○
			○
			○
8	9	10	○
			○
			○
			○
15	16	17	○
			○
			○
			○
			○
22	23	24	○
			○
			○
			○
			○
29	30		NOTES

December

MONDAY	TUESDAY	WEDNESDAY	THURSDAY
2	3	4	5
9	10	11	12
16	17	18	19
23	24	25	26

December

FRIDAY	SATURDAY	SUNDAY	NOTES
		1	○
			○
			○
			○
			○
6	7	8	○
			○
			○
			○
13	14	15	○
			○
			○
			○
			○
20	21	22	○
			○
			○
			○
			○
27	28	29	30 / 31

December
2023

01 FRIDAY

02 SATURDAY

03 SUNDAY

04 MONDAY

December
2023

05 TUESDAY

06 WEDNESDAY

07 THURSDAY

08 FRIDAY

December
2023

09 SATURDAY

10 SUNDAY

11 MONDAY

12 TUESDAY

13 WEDNESDAY

14 THURSDAY

15 FRIDAY

16 SATURDAY

December
2023

17 SUNDAY

18 MONDAY

19 TUESDAY

20 WEDNESDAY

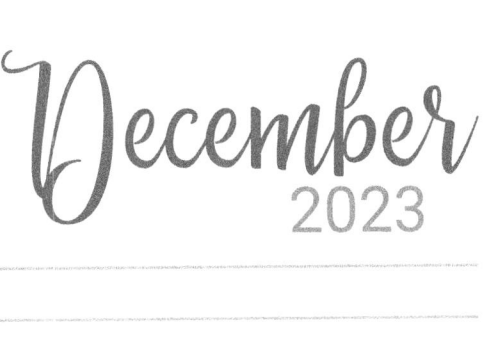

December
2023

21 THURSDAY

22 FRIDAY

23 SATURDAY

24 SUNDAY

25 MONDAY

26 TUESDAY

27 WEDNESDAY

28 THURSDAY

December
2023

29 FRIDAY

○
○
○
○
○
○
○
○

30 SATURDAY

○
○
○
○
○
○
○
○
○

31 SUNDAY

○
○
○
○
○
○
○
○

NOTES

January
2024

01 MONDAY
○
○
○
○
○
○
○
○

02 TUESDAY
○
○
○
○
○
○
○
○

03 WEDNESDAY
○
○
○
○
○
○
○
○

04 THURSDAY
○
○
○
○
○

January 2024

05 FRIDAY

○ _____
○ _____
○ _____
○ _____
○ _____
○ _____
○ _____
○ _____

06 SATURDAY

○ _____
○ _____
○ _____
○ _____
○ _____
○ _____
○ _____
○ _____

07 SUNDAY

○ _____
○ _____
○ _____
○ _____
○ _____
○ _____
○ _____
○ _____

08 MONDAY

○ _____
○ _____
○ _____
○ _____
○ _____

09 TUESDAY

○
○
○
○
○
○
○
○

10 WEDNESDAY

○
○
○
○
○
○
○
○

11 THURSDAY

○
○
○
○
○
○
○
○

12 FRIDAY

○
○
○
○
○

13 SATURDAY

○
○
○
○
○
○
○
○

14 SUNDAY

○
○
○
○
○
○
○
○

15 MONDAY

○
○
○
○
○
○
○
○

16 TUESDAY

○
○
○
○
○

17 WEDNESDAY

○
○
○
○
○
○
○

18 THURSDAY

○
○
○
○
○
○
○
○

19 FRIDAY

○
○
○
○
○
○
○
○

20 SATURDAY

○
○
○
○
○

21 SUNDAY

○
○
○
○
○
○
○
○

22 MONDAY

○
○
○
○
○
○
○
○

23 TUESDAY

○
○
○
○
○
○
○
○

24 WEDNESDAY

○
○
○
○
○

25 THURSDAY

26 FRIDAY

27 SATURDAY

28 SUNDAY

January 2024

29 MONDAY

30 TUESDAY

31 WEDNESDAY

NOTES

01 THURSDAY

02 FRIDAY

03 SATURDAY

04 SUNDAY

February
2024

05 MONDAY

○
○
○
○
○
○
○
○

06 TUESDAY

○
○
○
○
○
○
○
○

07 WEDNESDAY

○
○
○
○
○
○
○
○

08 THURSDAY

○
○
○
○
○

February
2024

09 FRIDAY

10 SATURDAY

11 SUNDAY

12 MONDAY

February
2024

13 TUESDAY

14 WEDNESDAY

15 THURSDAY

16 FRIDAY

17 SATURDAY

18 SUNDAY

19 MONDAY

20 TUESDAY

21 WEDNESDAY

22 THURSDAY

23 FRIDAY

24 SATURDAY

25 SUNDAY

26 MONDAY

27 TUESDAY

28 WEDNESDAY

29 THURSDAY

- ○
- ○
- ○
- ○
- ○
- ○
- ○
- ○

NOTES

March 2024

01 FRIDAY

○
○
○
○
○
○
○
○

02 SATURDAY

○
○
○
○
○
○
○
○

03 SUNDAY

○
○
○
○
○
○
○
○

04 MONDAY

○
○
○
○
○

05 TUESDAY

06 WEDNESDAY

07 THURSDAY

08 FRIDAY

March
2024

09 SATURDAY

10 SUNDAY

11 MONDAY

12 TUESDAY

13 WEDNESDAY

14 THURSDAY

15 FRIDAY

16 SATURDAY

17 SUNDAY

18 MONDAY

19 TUESDAY

20 WEDNESDAY

21 THURSDAY

○ _____
○ _____
○ _____
○ _____
○ _____
○ _____
○ _____
○ _____

22 FRIDAY

○ _____
○ _____
○ _____
○ _____
○ _____
○ _____
○ _____
○ _____

23 SATURDAY

○ _____
○ _____
○ _____
○ _____
○ _____
○ _____
○ _____
○ _____

24 SUNDAY

○ _____
○ _____
○ _____
○ _____
○ _____

March 2024

25 MONDAY

26 TUESDAY

27 WEDNESDAY

28 THURSDAY

29 FRIDAY

30 SATURDAY

31 SUNDAY

NOTES

April
2024

01 MONDAY

02 TUESDAY

03 WEDNESDAY

04 THURSDAY

05 FRIDAY

06 SATURDAY

07 SUNDAY

08 MONDAY

09 TUESDAY

10 WEDNESDAY

11 THURSDAY

12 FRIDAY

13 SATURDAY

14 SUNDAY

15 MONDAY

16 TUESDAY

17 WEDNESDAY

18 THURSDAY

19 FRIDAY

20 SATURDAY

April
2024

21 SUNDAY

22 MONDAY

23 TUESDAY

24 WEDNESDAY

25 THURSDAY

26 FRIDAY

27 SATURDAY

28 SUNDAY

April
2024

29 MONDAY

30 TUESDAY

NOTES

01 WEDNESDAY

02 THURSDAY

03 FRIDAY

04 SATURDAY

05 SUNDAY

○
○
○
○
○
○
○
○

06 MONDAY

○
○
○
○
○
○
○
○

07 TUESDAY

○
○
○
○
○
○
○
○

08 WEDNESDAY

○
○
○
○
○

May 2024

09 THURSDAY

10 FRIDAY

11 SATURDAY

12 SUNDAY

13 MONDAY

○
○
○
○
○
○
○
○

14 TUESDAY

○
○
○
○
○
○
○
○

15 WEDNESDAY

○
○
○
○
○
○
○
○

16 THURSDAY

○
○
○
○
○

17 FRIDAY

18 SATURDAY

19 SUNDAY

20 MONDAY

21 TUESDAY

○ _____
○ _____
○ _____
○ _____
○ _____
○ _____
○ _____
○ _____

22 WEDNESDAY

○ _____
○ _____
○ _____
○ _____
○ _____
○ _____
○ _____
○ _____

23 THURSDAY

○ _____
○ _____
○ _____
○ _____
○ _____
○ _____
○ _____
○ _____

24 FRIDAY

○ _____
○ _____
○ _____
○ _____
○ _____

25 SATURDAY

26 SUNDAY

27 MONDAY

28 TUESDAY

29 WEDNESDAY

○ _____
○ _____
○ _____
○ _____
○ _____
○ _____
○ _____
○ _____

30 THURSDAY

○ _____
○ _____
○ _____
○ _____
○ _____
○ _____
○ _____
○ _____

31 FRIDAY

○ _____
○ _____
○ _____
○ _____
○ _____
○ _____
○ _____
○ _____

NOTES

_____
_____
_____
_____

01 SATURDAY
○
○
○
○
○
○
○
○

02 SUNDAY
○
○
○
○
○
○
○
○

03 MONDAY
○
○
○
○
○
○
○
○

04 TUESDAY
○
○
○
○
○

05 WEDNESDAY

○
○
○
○
○
○
○

06 THURSDAY

○
○
○
○
○
○
○
○

07 FRIDAY

○
○
○
○
○
○
○
○

08 SATURDAY

○
○
○
○
○

09 SUNDAY

○
○
○
○
○
○
○

10 MONDAY

○
○
○
○
○
○
○
○

11 TUESDAY

○
○
○
○
○
○
○
○

12 WEDNESDAY

○
○
○
○
○

13 THURSDAY

14 FRIDAY

15 SATURDAY

16 SUNDAY

17 MONDAY

○
○
○
○
○
○
○

18 TUESDAY

○
○
○
○
○
○
○
○

19 WEDNESDAY

○
○
○
○
○
○
○
○

20 THURSDAY

○
○
○
○
○

21 FRIDAY

○
○
○
○
○
○
○
○

22 SATURDAY

○
○
○
○
○
○
○
○

23 SUNDAY

○
○
○
○
○
○
○
○

24 MONDAY

○
○
○
○
○

25 TUESDAY

○
○
○
○
○
○
○

26 WEDNESDAY

○
○
○
○
○
○
○
○

27 THURSDAY

○
○
○
○
○
○
○
○

28 FRIDAY

○
○
○
○
○

29 SATURDAY

○ _____
○ _____
○ _____
○ _____
○ _____
○ _____
○ _____

30 SUNDAY

○ _____
○ _____
○ _____
○ _____
○ _____
○ _____
○ _____
○ _____

NOTES

July
2024

01 MONDAY
- ○
- ○
- ○
- ○
- ○
- ○
- ○
- ○

02 TUESDAY
- ○
- ○
- ○
- ○
- ○
- ○
- ○
- ○

03 WEDNESDAY
- ○
- ○
- ○
- ○
- ○
- ○
- ○
- ○

04 THURSDAY
- ○
- ○
- ○
- ○
- ○

05 FRIDAY

○
○
○
○
○
○
○
○

06 SATURDAY

○
○
○
○
○
○
○
○
○

07 SUNDAY

○
○
○
○
○
○
○
○
○

08 MONDAY

○
○
○
○
○

09 TUESDAY

○
○
○
○
○
○
○

10 WEDNESDAY

○
○
○
○
○
○
○
○

11 THURSDAY

○
○
○
○
○
○
○
○

12 FRIDAY

○
○
○
○
○

13 SATURDAY

14 SUNDAY

15 MONDAY

16 TUESDAY

17 WEDNESDAY

18 THURSDAY

19 FRIDAY

20 SATURDAY

21 SUNDAY

22 MONDAY

23 TUESDAY

24 WEDNESDAY

25 THURSDAY

26 FRIDAY

27 SATURDAY

28 SUNDAY

July
2024

29 MONDAY

30 TUESDAY

31 WEDNESDAY

NOTES

August 2024

01 THURSDAY

○
○
○
○
○
○
○
○

02 FRIDAY

○
○
○
○
○
○
○
○

03 SATURDAY

○
○
○
○
○
○
○
○

04 SUNDAY

○
○
○
○
○

August 2024

05 MONDAY

06 TUESDAY

07 WEDNESDAY

08 THURSDAY

09 FRIDAY

10 SATURDAY

11 SUNDAY

12 MONDAY

13 TUESDAY

○ _____
○ _____
○ _____
○ _____
○ _____
○ _____
○ _____
○ _____

14 WEDNESDAY

○ _____
○ _____
○ _____
○ _____
○ _____
○ _____
○ _____
○ _____

15 THURSDAY

○ _____
○ _____
○ _____
○ _____
○ _____
○ _____
○ _____
○ _____

16 FRIDAY

○ _____
○ _____
○ _____
○ _____
○ _____

17 SATURDAY

18 SUNDAY

19 MONDAY

20 TUESDAY

August
2024

21 WEDNESDAY

○
○
○
○
○
○
○
○

22 THURSDAY

○
○
○
○
○
○
○
○

23 FRIDAY

○
○
○
○
○
○
○
○

24 SATURDAY

○
○
○
○
○

25 SUNDAY

26 MONDAY

27 TUESDAY

28 WEDNESDAY

August
2024

29 THURSDAY

○ _____
○ _____
○ _____
○ _____
○ _____
○ _____
○ _____

30 FRIDAY

○ _____
○ _____
○ _____
○ _____
○ _____
○ _____
○ _____

31 SATURDAY

○ _____
○ _____
○ _____
○ _____
○ _____
○ _____
○ _____

NOTES

_____
_____
_____

01 SUNDAY

○
○
○
○
○
○
○
○

02 MONDAY

○
○
○
○
○
○
○
○

03 TUESDAY

○
○
○
○
○
○
○
○

04 WEDNESDAY

○
○
○
○
○

September
2024

05 THURSDAY

06 FRIDAY

07 SATURDAY

08 SUNDAY

09 MONDAY

○
○
○
○
○
○
○
○

10 TUESDAY

○
○
○
○
○
○
○
○

11 WEDNESDAY

○
○
○
○
○
○
○
○

12 THURSDAY

○
○
○
○
○

13 FRIDAY

14 SATURDAY

15 SUNDAY

16 MONDAY

17 TUESDAY

○
○
○
○
○
○
○

18 WEDNESDAY

○
○
○
○
○
○
○
○

19 THURSDAY

○
○
○
○
○
○
○
○

20 FRIDAY

○
○
○
○
○

21 SATURDAY

○
○
○
○
○
○
○

22 SUNDAY

○
○
○
○
○
○
○

23 MONDAY

○
○
○
○
○
○
○
○

24 TUESDAY

○
○
○
○

25 WEDNESDAY

26 THURSDAY

27 FRIDAY

28 SATURDAY

29 SUNDAY

- ○
- ○
- ○
- ○
- ○
- ○
- ○
- ○

30 MONDAY

- ○
- ○
- ○
- ○
- ○
- ○
- ○
- ○

NOTES

October
2024

01 TUESDAY

02 WEDNESDAY

03 THURSDAY

04 FRIDAY

October
2024

05 SATURDAY

○
○
○
○
○
○
○
○

06 SUNDAY

○
○
○
○
○
○
○
○

07 MONDAY

○
○
○
○
○
○
○
○

08 TUESDAY

○
○
○
○
○

09 WEDNESDAY

10 THURSDAY

11 FRIDAY

12 SATURDAY

13 SUNDAY

14 MONDAY

15 TUESDAY

16 WEDNESDAY

17 THURSDAY

○
○
○
○
○
○
○
○

18 FRIDAY

○
○
○
○
○
○
○
○

19 SATURDAY

○
○
○
○
○
○
○
○

20 SUNDAY

○
○
○
○
○

October
2024

21 MONDAY

22 TUESDAY

23 WEDNESDAY

24 THURSDAY

25 FRIDAY

26 SATURDAY

27 SUNDAY

28 MONDAY

October
2024

29 TUESDAY

○
○
○
○
○
○
○
○

30 WEDNESDAY

○
○
○
○
○
○
○
○

31 THURSDAY

○
○
○
○
○
○
○
○

NOTES

01 FRIDAY

02 SATURDAY

03 SUNDAY

04 MONDAY

05 TUESDAY

06 WEDNESDAY

07 THURSDAY

08 FRIDAY

09 SATURDAY

○
○
○
○
○
○
○

10 SUNDAY

○
○
○
○
○
○
○
○

11 MONDAY

○
○
○
○
○
○
○
○

12 TUESDAY

○
○
○
○
○

13 WEDNESDAY

14 THURSDAY

15 FRIDAY

16 SATURDAY

17 SUNDAY

18 MONDAY

19 TUESDAY

20 WEDNESDAY

November
2024

21 THURSDAY

22 FRIDAY

23 SATURDAY

24 SUNDAY

25 MONDAY

26 TUESDAY

27 WEDNESDAY

28 THURSDAY

November
2024

29 FRIDAY

○ _____
○ _____
○ _____
○ _____
○ _____
○ _____
○ _____
○ _____

30 SATURDAY

○ _____
○ _____
○ _____
○ _____
○ _____
○ _____
○ _____
○ _____

NOTES

December
2024

01 SUNDAY

02 MONDAY

03 TUESDAY

04 WEDNESDAY

December
2024

05 THURSDAY

06 FRIDAY

07 SATURDAY

08 SUNDAY

December
2024

09 MONDAY

10 TUESDAY

11 WEDNESDAY

12 THURSDAY

December
2024

13 FRIDAY
- ○
- ○
- ○
- ○
- ○
- ○
- ○
- ○

14 SATURDAY
- ○
- ○
- ○
- ○
- ○
- ○
- ○
- ○

15 SUNDAY
- ○
- ○
- ○
- ○
- ○
- ○
- ○
- ○

16 MONDAY
- ○
- ○
- ○
- ○
- ○

December
2024

17 TUESDAY

18 WEDNESDAY

19 THURSDAY

20 FRIDAY

21 SATURDAY

○
○
○
○
○
○
○
○

22 SUNDAY

○
○
○
○
○
○
○
○

23 MONDAY

○
○
○
○
○
○
○
○

24 TUESDAY

○
○
○
○
○

25 WEDNESDAY

26 THURSDAY

27 FRIDAY

28 SATURDAY

December
2024

29 SUNDAY

○
○
○
○
○
○
○
○

30 MONDAY

○
○
○
○
○
○
○
○

31 TUESDAY

○
○
○
○
○
○
○
○

NOTES